GREAT APES

Protecting Our Animal Cousins

CHRISTOPHER GUDGEON

ORCA BOOK PUBLISHERS

Published in Canada and the United States in 2025 by Orca Book Publishers.
orcabook.com

Library and Archives Canada Cataloguing in Publication
Title: Great apes : protecting our animal cousins / Christopher Gudgeon.
Names: Gudgeon, Christopher, 1959- author.
Series: Orca wild.
Description: Series statement: Orca wild ; 15 | Includes bibliographical references and index.
Identifiers: Canadiana (print) 20240349040 | Canadiana (ebook) 20240349172 | ISBN 9781459838109 (hardcover) | ISBN 9781459838116 (PDF) | ISBN 9781459838123 (EPUB)
Subjects: LCSH: Apes—Juvenile literature. | LCSH: Apes—Conservation—Juvenile literature.
Classification: LCC QL737.P94 G83 2025 | DDC j599.88—dc23

Library of Congress Control Number: 2024936047

Summary: This nonfiction book introduces middle grade readers to great apes, including chimpanzees, gorillas, orangutans and bonobos. Featuring photos throughout, it explores great ape habitats, biology, threats to survival and how conservationists, scientists and young people are working to protect them.

Orca Book Publishers is committed to reducing the consumption of nonrenewable resources in the production of our books. We make every effort to use materials that support a sustainable future.

Orca Book Publishers gratefully acknowledges the support for its publishing programs provided by the following agencies: the Government of Canada, the Canada Council for the Arts and the Province of British Columbia through the BC Arts Council and the Book Publishing Tax Credit.

Front cover photo by Anup Shah /Getty Images.
Back cover photo by Freder / Getty Images.
Design by Troy Cunningham.
Edited by Kirstie Hudson.

Printed and bound in South Korea.

28 27 26 25 • 1 2 3 4

Chimpanzees like this one are part of the great ape family.
YANNICK TYLLE/GETTY IMAGES

To George, who's still too young to read this. Let's hope there's still lots of great apes in the wild by the time you're old enough to go and visit them.

Contents

4

5

6

Bonobos are one of our closest relatives in the animal kingdom.
FIONA ROGERS/GETTY IMAGES

Introduction
Molly and Me

Many years ago, when I was still a kid, I went to a small private zoo with my family. The zoo had a bunch of different animals, all of them locked up in concrete and metal cages. It was located in the mountains of British Columbia, and most of the animals were from the general area. There were rattlesnakes, a couple of coyotes, some cougars and a very old grizzly bear who just lay on the floor of his cage, picking off bits of his fur. All the animals seemed unhappy and bored, locked up like that all day long. It made me sad to see them like that.

One animal really stood out, though. I noticed her as soon as our car pulled into the zoo parking lot. She was sitting on a tire swing suspended from the ceiling of her cage, slowly swaying back and forth. At first I thought it was a chimpanzee and wondered, What is one of *those* doing way up here in the mountains of Canada? Once I got up to the cage, I realized it wasn't a chimp at all. I read the handwritten sign screwed onto the bars. It said *Molly, a female bonobo*.

Bonobo? I'd never heard of that.

It turns out they're close relatives of the chimp. They look so similar, in fact, that it took scientists decades to realize they were actually two different ***species***. Chimp or bonobo, I was excited to see a real ape close up. I thought Molly would be just like the apes I'd seen on TV, a happy, playful almost-human. I thought she'd be excited to see me too. I was a nice, friendly kid. Maybe we could play with each other, even though we were on other sides of the cold steel bars.

A BETTER, HAPPIER FUTURE

Molly had no interest in me. She was just as sad and bored as all the other animals. As I watched her dangling aimlessly from the worn-out tire, I had a strong urge to open her cage and set her free. I didn't, of course. I wasn't that kind of kid. Besides, it probably would have made things worse for Molly. She was from the jungles of Africa. She would have had a hard time surviving in those cold, harsh North American mountains.

She stuck in my mind, though. When I got home, I went to the library. I researched this curious new creature, the bonobo. From there, I started reading up on other great apes—gorillas, orangutans…and, of course, chimpanzees.

I already knew that when I grew up, I wanted to be an author. I promised myself that one day I would write about Molly and all the other great apes of the world. I wanted to share all the interesting things I had learned and, hopefully, help great apes have a better, happier future.

At long last, I've kept my promise. *Great Apes: Protecting Our Animal Cousins* is your introduction to these amazing animals. Starting with the smallest—bonobos like Molly—and working up to the largest—the goliath gorillas—it's a guided tour through the great ape world.

I hope you find them as fascinating as I do, and maybe it will spark something in you too. We share our planet with these remarkable apes, but so many of them are in danger today. Let's discover more about them, and along the way maybe I'll inspire you to try to help them. I'm sure that would make Molly very happy.

We share our planet with remarkable apes like this majestic silverback gorilla.
ANUP SHAH/GETTY IMAGES

The ape family includes lesser apes like this lar gibbon.

BILLY CURRIE PHOTOGRAPHY/GETTY IMAGES

1

Planet of the (Great) Apes

They are some of the most remarkable animals on Earth. Living in the lush rainforests of Southeast Asia and the dense jungles and ***savannas*** of Africa, their intelligence, social structures and communication skills set them apart from other animals. Bonobos, chimpanzees, gorillas, orangutans...these are the great apes, and they are our closest relatives in the animal kingdom. What's so great about them? It's not just that they have wonderful personalities. They are big, as in *great* big. Scientists use the term *great ape* to separate them from the 20 or so members of the lesser ape family. These are the tree-dwelling gibbons that live in the rainforests of Bangladesh, India, China, Cambodia, Thailand and Indonesia. They are *lesser* because their bodies and brains are smaller. Great apes, big. Lesser apes, small. Got it?

Siamangs are the largest member of the gibbon family.
STEVE CLANCY PHOTOGRAPHY/GETTY IMAGES

LESSER IS MORE

While there are four kinds of great apes, there is only one lesser ape, the gibbon. Gibbons live in tropical and subtropical forests of Southeast Asia.

The gibbon family includes 20 separate species, ranging from the largest, the siamangs, which reach almost 3 feet (91 centimeters) in length and weigh up to 22 pounds

CHARLES DARWIN, EVOLUTION'S DARLING

"It is not the most intellectual of the species that survives; it is not the strongest that survives; but the species that survives is the one that is able best to adapt and adjust to the changing environment in which it finds itself."

—CHARLES DARWIN

Charles Darwin is one of the most important scientists who ever lived. He was born over 200 years ago in England. As a young man he took part in a scientific expedition that went around the world on a ship called the HMS ***Beagle***, exploring unique places and the plants and animals that lived in them. One thing Darwin noticed was that while certain animals might look similar to their relatives in another part of the world, they often had unique features. These features allowed them to deal with the specific challenges found in their specific worlds.

ALL IN THE GIBBON FAMILY

Gibbons live in small family groups, kind of like us humans, and are very sociable. They love hanging out—literally—with their extended kin. But don't push them too far. Gibbons are very ***territorial***, and they will attack other animals, including other gibbons and even the occasional human, if they think their family or home is being threatened.

(10 kilograms)—about the same as a large watermelon—to smaller species like the lar and the black-handed gibbon, which grow up to 23 inches (58 centimeters) and weigh a maximum of 15 pounds (6.8 kilograms).

Regardless of their size, all gibbons are amazing aerial acrobats. With their long arms and strong hands, they can swing through the treetops at 34 miles per hour (55 kilometers per hour) and can travel almost 50 feet (15 meters) in a single swing. Gibbons spend their entire lives in the trees and have evolved in ways that help them get by in their ***arboreal*** world. They have very long fingers and toes, which allow them to grab a nearby branch and hold on tight. Unlike monkeys, though, and like other apes, gibbons do not have tails.

Gibbons are expert tree climbers and can travel almost 50 feet (15 meters) in a single swing.

AYIMAGES/GETTY IMAGES

For example, on one Galapagos island he found a species of finch that was closely related to similar birds he'd seen elsewhere in the world. But the Galapagos finch had a long, narrow beak, perfectly designed to sip nectar out of the elongated ***stamen*** of its favorite flower. This helped Darwin develop his theory of ***evolution*** a groundbreaking idea that explained how all living things—past, present and future—are linked together through time.

FRANCO TOLLARDO/GETTY IMAGES

EVOLUTION ANYONE?

Evolution is an important scientific idea that helps us understand how living things change over time. The theory also shows us that all things—plants and animals and even humans—are connected. Behind the idea is the understanding that Earth's geology and environment are constantly transforming. In order for groups of living things to survive, they need a strategy to adapt as their world shifts.

Key to evolution is something called natural selection. This means that nature helps decide how a species gradually changes. Even though all the members of a species are similar, every individual is slightly different. Think of all the kids in your classroom. You're all humans, studying the same things, playing together at recess. But you all have unique traits. These differences are nature's way of making sure we have lots of options to help us as a species adapt to new features in our world.

For example, have you ever wondered why giraffes have long necks? Giraffes love to eat the branches and leaves of acacia, mimosa and other wild fruit trees. A long time ago, the ancestors of giraffes lived in a place where there were lots of these trees. The longer your neck was, the better able you were to reach the tasty bits way up high. And those animals who could reach these bits became bigger and stronger...and made more attractive mates. In the giraffe world, the longer your neck is, the hotter you are. It took thousands of years for the stubby-necked giraffe ancestors to evolve into our familiar long-necked friends, and all because of the evolutionary process called natural selection.

The giraffe has evolved to have a long neck so it can reach high up into the trees to eat its favorite foods.
CLARA MANOLACHE/GETTY IMAGES

These curious creatures also play a critical role in the forest ***ecosystem***. As they range through the treetops, grazing on figs, liana fruit and berries, they poop out seeds that land on the forest floor. Gibbon poop is a perfect fertilizer, helping the seeds ***germinate*** and eventually grow into new fruit trees.

WHAT MAKES AN APE AN APE?

Do you think apes are a kind of monkey? Well, think again. They are a different kind of animal altogether. But don't worry. Lots of people think apes and monkeys are the same thing. How can you tell apes apart from other ***primates***

MONKEYS AND APES, RELATIVELY SPEAKING

Turns out, apes and monkeys share 95 percent of their ***genes***. That sounds like a lot until you realize that we share more than 98 percent of our ***DNA*** with gorillas and a whopping 99 percent with chimpanzees and bonobos. This is important because the more DNA two species have in common, the more closely related they are.

Lar gibbons and other apes share 95 percent of their DNA with monkeys.
BORCHEE/GETTY IMAGES

like monkeys and lemurs? Apes are bigger, stronger and don't have tails. They do have something monkeys don't, though—***opposable thumbs***. This means apes can pick up things like sticks and rocks and use them as tools.

Another thing about apes. While all primates are smart, apes have *big* brains compared to their body size. This helps them problem-solve, communicate with one another and live in complex social groups. Apes can also walk on two feet. Monkeys can't. True, most apes prefer to walk on all fours. But all apes are able to do it. These are some of the most obvious differences. But look closer, and the distinctions are even clearer.

Take ape skeletons. You'd think apes and monkeys would have similar bone structures. But underneath it all, monkeys look more like dogs, while apes look more like humans. Look even closer. In fact, look as close as you can get, right down to the ***cells***. Using powerful microscopes, scientists are able to study great ape DNA and compare it to the DNA of other primates.

A FAMILY REUNION?

Hang on a second.

No tail?

Opposable thumbs?

Big brains?

Does that mean we're apes too?

It sure does.

In the long history of life on this planet, humans and great apes are about as similar as you can get. In fact, they're our closest relatives in the animal kingdom.

HIMPANZEES? GIRLILLAS?

Given how closely related we are, is an ape-human ***hybrid*** possible? Believe it or not, some scientists have studied this question. In fact, there are rumors that scientists in Russia, China and the United States have actually tried to crossbreed humans with apes. If these experiments ever took place, the results remain top secret. In theory, though, it is possible. Apes and humans share a lot of DNA and are more closely related than some other species that have successfully interbred.

Mules are the best-known hybrid. The offspring of a female horse and a male donkey, mules are bigger and stronger than their fathers, and while they can't run as fast as their moms, they can go much farther. (Impress-your-friends bonus: the offspring of a male horse and a female donkey is called a hinny!). Less well known are zorses, which have a zebra father and a horse mother and usually look like a smaller, zebra-striped horse. People have also created ligers, (the offspring of lions and tigers), geeps (goats and sheep) and grolar bears (yikes!).

Have you heard the expression ***family tree***? It's a way of using a picture to help show how you and all your relatives fit together. The roots of the tree might represent your parents and grandparents. The branches nearest to you are your brothers and sisters. The branches a little farther away are your cousins and cousins' cousins and cousins' cousins' cousins' cousins.

Animals have family trees too. It's how scientists track the relationship between different animals (and plants!). The canine family tree includes dogs, wolves, dingoes and foxes. In the feline family, you find cats, lions, tigers, cheetahs and cougars. The ape family tree is where gibbons, gorillas (and other great apes) and humans hang out. How do scientists decide what family an animal belongs to? One clue is to look for physical similarities, things like size, shape, color, the structure of the skeleton and so on. It's easy to tell just by looking at a dog, for example, that it must be related to a wolf.

Sometimes it's harder to figure out where an animal belongs. It took centuries for humans to realize that they are part of a family that includes chimps and gorillas. Scientific advances have made it easier to figure out who belongs where.

So close, in fact, that scientists classify both humans and great apes as ***hominoids.*** This doesn't mean that chimps and orangutans are our great-great-great-great grandparents. But we do share a common ancestor.

Millions and millions of years after the so-called ***Age of Dinosaurs,*** a tiny creature lived in the jungles of what we now call Africa. We don't know what this animal looked like. But we do know that it had traits in common with both great apes and humans. Over time this creature changed into different species, each one suited to its own unique environment. Great apes, for example, moved into the forests and became climbers and plant-eaters. Humans took another path. We started walking on two legs, which gave us an advantage in ***grasslands***, savannas and plains.

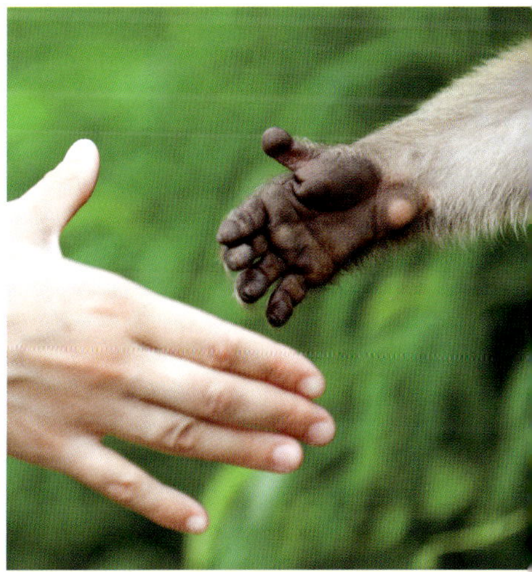

Both humans and apes have opposable thumbs, which makes it easier to hold things and use tools.
LAOSHI/GETTY IMAGES

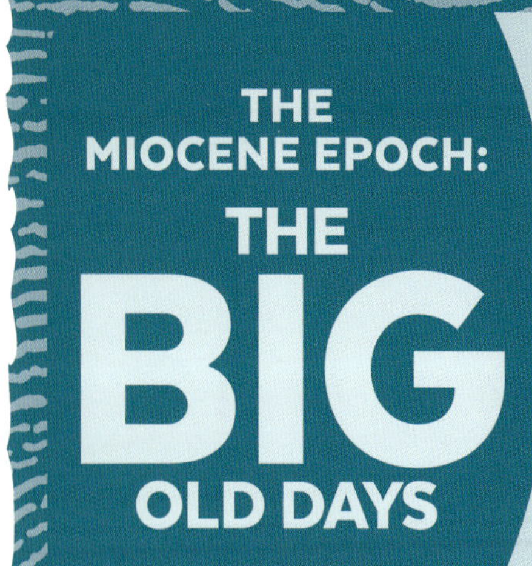

Because of its unique climate, the Miocene epoch produced a lot of very big animals and plants. Enormous ground sloths, massive rhinoceroses and colossal land-dwelling birds were among the incredible animals that roamed Earth during this time. It was also when modern mammal groups began to emerge...including early apelike creatures. In fact, Earth really was the planet of the apes back then. There were more than 100 ape species spread out across what is now Europe, Africa, Asia and China. These early apes came in all different shapes and sizes, from the ***Similous minutus*** ("miniature primate"), which was no larger than a house cat, to the ***Gigantopithecus***, which stood 10 feet (3 meters) tall.

GREAT APES ON THE EDGE

These days great apes are struggling. Because they live in remote areas and spend most of their time in tree-tops and dense forests, it's hard for scientists to keep track of them. Our best guess is that there are between 500,000 and 700,000 great apes living in the wild today.

Way back in 1935, a Dutch anthropologist named Gustav Ralph von Koenigswald stumbled across a remarkable find in a Hong Kong market. Tucked away on the top shelf of a traditional drugstore was a glass jar containing what were labeled ***dragon bones***. But Koenigswald had a hunch, and after a a little research he identified them as teeth from some kind of long-lost and very large primate. Koenigswald named the beast ***Gigantopithecus***, a made-up Greek word that means "giant ape."

Over time scientists discovered more teeth, along with hundreds of bone fragments, all belonging to this

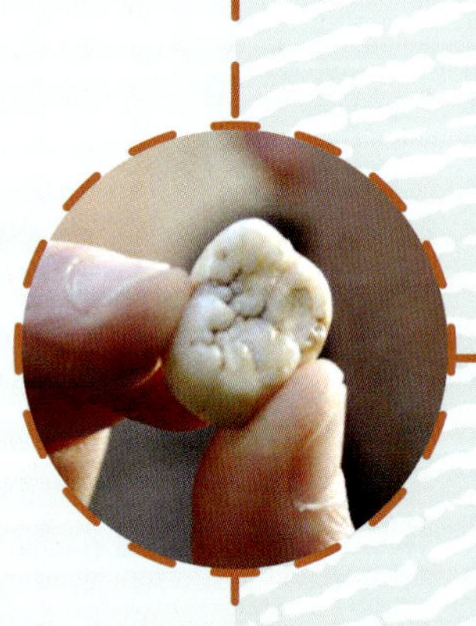

GERBIL/WIKIMEDIA COMMONS/CC BY-SA 3.0 DEED

Twenty years ago, there were four or five million of them.

These days you find great apes only in remote jungles and mountain forests in limited areas of Africa and Asia. But once upon a time, apes were almost everywhere. It was during the Miocene, an epoch that lasted from 23 to 5 million years ago. The Miocene had a relatively stable climate. Earth was going through a long warm period when temperatures were generally higher than they are now. Much of the land was covered in lush, dense forests. Swamps, lakes and rivers were abundant.

But the glory days for apes are long gone. Great ape populations in Africa and Asia have gone way down over the past 30 years because of things like ***habitat*** loss, illegal hunting and disease.

Great ape populations are declining as their forest homes shrink.

CURIOUSTIGER/GETTY IMAGES

mysterious, long-gone creature. With these bits and pieces—and the help of some powerful computers—we now have a pretty good of idea of what *Gigantopithecus* looked like. It was big, and by *big*, we mean ***BIG!*** Imagine a orangutan 10 feet (3 meters) tall—that's as high as a basketball hoop and twice the size of a full-grown eastern lowland gorilla! With arms so long they could reach all the way around a small car, this big boy weighed in at a whopping 600 pounds (272 kilograms).

As big and scary-looking as it might have been, *Gigantopithecus*—like some great apes today—was a vegetarian, with big, strong jaws perfectly suited for munching branches, thick leaves and bark. *Gigantopithecus* thrived in the **semitropical** forests of China and Southeast Asia for almost two million years. Pretty impressive considering we humans have only been around for about 300,000 years. About 100,000 ***BCE***, these great, great apes suddenly vanished. No one knows exactly why they went ***extinct***, but it probably had something to do with climate change. As the world got colder and plants got smaller, *Gigantopithecus* was probably just too gigantic to survive.

Recent studies predict that if things don't change quickly, the great apes' habitat will be gone within 20 years.

But it's not all bad news. Conservation efforts are making a difference. The first step is education. The more we know about the amazing great apes, the more we can do to help them. Let's get to know them better and explore the world—and life—of our closest animal cousins.

THE MISSING LINK

An important piece of the evolution puzzle is what's called the missing link. That's an ancient creature that is the ancestor of two very different modern animals. For example, ***Tiktaalik*** is a missing link between fish and land-based animals that lived around 375 million years ago. These ancient creatures lived in water and had scales and gills. But they also had strong bones in their bottom fins that helped them move around on land. Then there's ***Archaeopteryx***, which lived about 150 million years ago. These curious creatures were about the size of a raven and had wings and feathers, like a modern bird. But they didn't have beaks. They had strong jaws with teeth, not to mention three clawed fingers, which are the kinds of things you'd normally find in a dinosaur.

While scientist haven't found any fossils of the missing link between humans and apes, they are getting closer. Ardi (short for ***Ardipithecus ramidus***) lived 4.4 million years ago. Ardi's bones tell us it could walk on two legs like a human but also had long arms perfectly suited for life in the trees. Then there's Lucy (or ***Australopithecus afarensis***). She lived around 3 million years ago and had a curious mix of human and ape traits. Her skull was small and her arms long, and she had a cone-shaped rib cage, all features of apes today. But her spine was long and straight, and her pelvis and knees indicate that she walked upright, just like we do today. Discoveries like Ardi and Lucy are important clues. Who knows? Maybe one day you'll be the scientist who makes the important discovery of the ape-human missing link.

(TOP) NOBUMICHI TAMURA/STOCKTREK IMAGES/GETTY IMAGES; (BOTTOM) TIM BOYLE/GETTY IMAGES

KNOW YOUR GREAT APES

Here's a quick guide to the four kinds of great apes.

1 **GORILLAS.** Large and strong, gorillas have black fur and are known for their peaceful nature.

2 **CHIMPANZEES.** These playful apes have brown fur and are excellent climbers and problem-solvers.

3 **BONOBOS.** Similar to chimpanzees but slimmer, bonobos are known for their peaceful nature and social behaviors.

4 **ORANGUTANS.** These orange-furred apes are the only great apes found in Asia. They are great tree climbers.

(TOP) VICKI JAURON, BABYLON AND BEYOND PHOTOGRAPHY/GETTY IMAGES; (BOTTOM) PAULA BRONSTEIN/GETTY IMAGES

Bonobos, the smallest of the great apes, are the happy hippies of the Congo forests.

ANUP SHAH/GETTY IMAGES

2

Bonobos, Nature's Happy Hippies

NEWEST MEMBER OF THE GREAT APE FAMILY

In the rainforests of the Congo basin live the smallest of the great apes, the remarkable bonobos. Gentle, fun-loving and very social, they are described by some people as the happy hippies of the animal kingdom. Bonobos are primarily arboreal. That means they spend most of their lives in trees. Their long arms and flexible shoulders make it easy for them to swing through the ***forest canopy***, where these ***herbivores*** can find a rich supply of their favorite foods, such as fruits, leaves and other vegetation.

For centuries scientists thought bonobos were a kind of scaled-down chimpanzee. In fact, people called them pygmy chimps. In the 1920s scientists took a close look at bonobos and realized they were a separate species with unique characteristics and behaviors.

FAST FACTS

NAME ORIGIN: Might come from Bolobo, the traditional name for the area in Africa where scientists first studied these apes.

SCIENTIFIC NAME: *Pan paniscus*. *Pan* is the Greek word for the genus, or group, to which bonobos and chimpanzees belong. *Paniscus*, means "little Pan"—is the name of the species (a group of animals within a specific genus). The *Pan* genus also includes chimpanzees (*Pan troglodytes*).

SUBSPECIES: None

SIZE: Both male and female bonobos stand 3.8 feet (1.1 meters) tall. Males weigh about 85 pounds (39 kilograms), females 68 pounds (31 kilograms).

RANGE: Found only in a 77,220-square-mile (200,000-square-kilometer) rainforest on the southern tip of the Democratic Republic of the Congo.

HABITAT: Arboreal.

FOOD: Bonobos are herbivores who love fruit but also eat leaves, flowers, seeds, bark, fungus and honey.

SOCIAL LIFE: They live in groups of 30 or more family members headed by a dominant female.

FAMILY LIFE: Bonobo moms have their first babies in their early teens.

LIFE SPAN: 45 to 50 years in the wild.

POPULATION: 15,000 to 20,000.

CONSERVATION STATUS: Endangered. This means they are at risk of becoming extinct.

MAIN THREATS: ***Deforestation***, illegal hunting.

It's easy to see why people were confused for so long. Bonobos and chimps share 99.6 percent of their DNA. That's about as close as two species can get.

TWO SPECIES ARE BETTER THAN ONE

Two species, so similar and yet very different? How did that happen? The answer is evolution. Bonobos and chimps share a common ancestor that lived a couple of million years ago. Over time the Congo River formed, separating a small group of these ancient apes. As the next million years or so passed, the two groups evolved into separate species, similar in many ways but different in others.

One big influence was environment. Living in a dense forest roughly the size of Spain, separated on all sides by rivers, bonobos had to cooperate in order to best use their resources. Being so close to the water, bonobos had easy access to plenty of food. This less-stressful lifestyle shaped their bodies. They didn't need to be big and aggressive to survive. That's why they are smaller and not as strong as their chimp cousins.

Bonobos love to dance...and they sure have some cool moves. They swing their arms, hop around and even spin in circles. Their dance parties aren't just for fun—they also help strengthen their bonds with one another.

PARTY ANIMALS!

GIRL POWER!

It's not just the physical differences that set bonobos apart. When it comes to their social behavior, the two species could not be more different. Chimps live in large groups headed by dominant males and are known to be very territorial. If other chimps try to move into their space, look out. They could very well attack...and full-scale chimp wars have been known to break out.

Bonobos live in smaller family groups, headed by strong female ***matriarchs.*** This is unusual for primates, which usually live in groups led by a forceful male. Female bonobos form bonds with one another and display exceptional communication skills, using various sounds, gestures and facial expressions to signal their emotions and intentions. When conflicts arise, bonobos don't get angry. Instead they will hug it out or groom one another.

The loving and caring nature of bonobos is also evident in the way they raise their young. Bonobo moms are the best, providing their kids with constant care and protection. And they are always willing to lend a hand to other moms. Being a young bonobo is kind of like being in one big, happy family with lots of caring aunties around. The result is that bonobos grow up to be caring adults.

Bonobos would rather hug it out than fight.
FIONA ROGERS/GETTY IMAGES

Bonobo moms are the best.
MARTIN HARVEY/GETTY IMAGES

They have remarkable ***empathy*** and know when a troopmate is sad or distressed. Feeling down in the dumps? Rest assured, one of your bonobo besties will swing by to cheer you up.

BRINGING UP BABY

Baby bonobos are born with light-pink faces that darken as they grow. They are supercute and rely on their moms for care and protection. Young bonobos learn a lot from their families, like what foods are good to eat and how to communicate with one another.

Baby bonobos, like all great apes, take several years to grow up. During the first year of life, a baby bonobo relies heavily on its mother for care and protection. The mother

AN UNSPOKEN CONNECTION

Scientists have noticed how similar bonobos are to human children. You often see them playing tag, play fighting, tickling each other and just hanging out, enjoying each other's company. This empathy goes beyond their troopmates. Bonobos in captivity can understand and respond to human gestures, spoken requests and even emotions. Some bonobos have been taught sign language to communicate and share their thoughts and feelings with humans.

Bonobo children love to eat, play tag and wrestle, just like you!
ANUP SHAH/GETTY IMAGES

nurses the infant, carries it on her back and provides for all its needs. The infant learns to hold on to its mother and gradually gains strength and coordination.

As the baby bonobo grows, it becomes more active and starts to explore its surroundings. It still stays close to its mother and learns from observing her actions and interactions with others. The mother continues to provide guidance and protection. After five or six years, a bonobo starts to become more independent and interact more with other members of its community. It's also more capable of finding food but still nurses occasionally and seeks comfort from its mom.

BOUNDING BONOBOS

Bonobos aren't just the nice guys of the great ape family but are also the best jumpers. An adult bonobo can leap almost 30 inches (76 centimeters) into the air—a useful trick if it encounters a hungry crocodile hiding in the Congo River reeds. By comparison, the leap of your average human—who's about twice the size of a bonobo—tops out at about 24 inches (61 centimeters).

TROUBLE IN BONOBO PARADISE

Playful, cooperative, friendly...it sounds like bonobos live in an ideal world. But, sadly, they face many threats, and their population has been declining for 30 years. Today there are fewer than 20,000 bonobos in the wild. You could seat them all in an average football stadium. Things are getting so serious that the ***United Nations*** has classified bonobos as critically endangered. That means they face a very high risk of becoming extinct in the near future.

Why are bonobos in so much trouble? The biggest problem is that their habitat is shrinking. Logging and farming are decreasing the size of the rainforest. Also, the Congo basin is still feeling the effects of a civil war that raged through the region at the end of the last century. The war made life hard for a lot of people. Food became scarce. Before the war, traditional ***taboos*** prevented locals from eating bonobos. But, faced with growing poverty

The more people log, the less rainforest habitat there is for bonobos.
PER-ANDERS PETTERSSON/GETTY IMAGES

and hunger, people are ignoring the taboos and hunting bonobos for food.

It's not all bad news. Many people in the Congo are fighting back and doing their best to protect bonobos. Organizations like the Bonobo Conservation Initiative are raising money to protect these gentle apes and their habitat.

Conservation groups are working to protect bonobos and their habitat.
MARTIN HARVEY/GETTY IMAGES

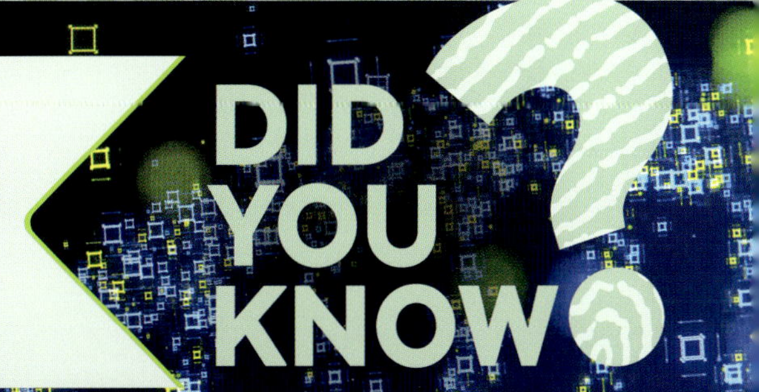

Here's something weird. Humans share a small part of their DNA (less than 2 percent) with bonobos but not with chimps. Likewise, we share a similar amount of different DNA with chimps but not with bonobos. This means that, in a small way, we are more closely related to these two apes than they are to each other.

Kanzi has a "conversation" with one of his caretakers.

MEET KANZI, THE SMARTEST (NONHUMAN) APE IN THE WORLD

Have you ever heard of Kanzi? He might just be the Albert Einstein of the animal world. Kanzi lives at a research facility in the United States called the Ape Cognition and Conservation Initiative. It's a place where scientists study how animals like Kanzi think.

Since coming to the facility, Kanzi has learned to communicate with humans using something called a lexigram. It's a kind of keyboard that uses little pictures instead of letters. Each picture represents a different word. To get his ideas across, Kanzi points to a series of pictures. It's like talking in sentences but using symbols instead of words. Kanzi has an everyday vocabulary of 450 words (including ***tummy***, ***tickle***, ***banana*** and ***fire***). His caretakers say he can understand 1,000 other words. That's about the same as an average four-year-old human. Not too shabby!

Kanzi isn't just good with words. He's also great at problem-solving. He loves searching for hidden treats and doing simple tasks his caretakers ask him to do. If you say, "Hey, Kanzi, go get your blanket," he's off to get it in a flash. One time Kanzi even used a stick to get a treat that was hard to reach. No one showed him how to do this. That means he is able to come up with creative ways to solve problems completely on his own. Scientists have learned a lot by watching Kanzi at work and play. His amazing abilities are a reminder that all animals are smart in their own ways. We have as much to learn about animal friends like Kanzi as they have to learn from us!

FROM PYGMY CHIMP TO BONOBO

1900s: Scientists believed that bonobos were part of the chimp species. Both apes lived in the dense forests of Africa and looked quite similar.

1920s: A zoologist named Harold Coolidge noticed that the chimps in a specific part of the Congo basin were distinct from other populations. Not only were these apes slimmer, with shorter faces and distinctive pink lips, but they also behaved differently from other chimps. They were more peaceful, less territorial and lived in groups headed by dominant females.

1950s: Biologists decided to classify these kinder, gentler apes as a ***subspecies*** of chimps, dubbing them bonobos (*bonobo* was probably based on the misspelling of Bolobo, a town on the Congo River near where these animals lived).

1984: The world was introduced to a new species, *Pan paniscus*—the scientific name given to bonobos.

Pink lips and black ears help set bonobos apart from chimpanzees.
ANUP SHAH/GETTY IMAGES

Chimps are the cool kids of the forests.
ANUP SHAH/GETTY IMAGES

3

Chimpanzees, the Cool Kids on the Ape Family Tree

CHIMPS: OUR COOL, COMPLEX COUSINS

In the heart of African forests lives a remarkable kind of primate known as the chimpanzee—chimp, for short. These amazing creatures are one of our closest relatives in the animal kingdom, sharing many similarities with humans. With their complex social lives and uncanny intelligence, chimps are a favorite of scientists and kids around the world. One thing you can say about chimps: they sure know how to have fun. With their long arms and strong muscles, they love to swing through the branches of the lush African woodlands they call home.

We've already seen how chimps and bonobos share a common ape ancestor. Some of the descendants of this ape wound up in an isolated river basin in the Congo, with lots of food and few predators. They evolved into a new arboreal species, the bonobos. The rest of this

FAST FACTS

NAME ORIGIN: From a similar-sounding word in the Bantu language meaning "ape" or "human animal."

SCIENTIFIC NAME: *Pan troglodytes. Pan* is the Greek word for members of the bonobo and chimp genus, or group. *Troglodytes* means "cave dweller" (some chimps live in caves part of the time).

SUBSPECIES: There are four recognized chimp subspecies: western, central, eastern and Nigeria-Cameroon. Each group lives in a specific area and has many unique traits.

SIZE: Chimps can grow to almost 5 feet (1.5 meters) tall. While males can weigh up to 150 pounds (70 kilograms), females max out at around 110 pounds (50 kilograms).

RANGE: Found in 21 countries across Central and West Africa.

HABITAT: Mostly arboreal, but chimps can often be found on the ground and even living in caves.

FOOD: Chimps are ***omnivorous***. Mostly they eat plants, roots and nuts, but they love fruit most of all. They'll also eat insects and small animals, but meat makes up only a tiny part of their diet.

SOCIAL LIFE: Chimp family groups can have as many as 150 members and are headed by dominant males called silverbacks.

FAMILY LIFE: Chimp moms have their first babies at around age 13. They have one baby at a time and can give birth every five years.

LIFE SPAN: Chimps in the wild typically make it to their 30th birthday, but some have been known to live for more than 60 years.

POPULATION: 170,000 to 300,000 in the wild. A hundred years ago, there were one to two million chimps.

CONSERVATION STATUS: All four subspecies are endangered and at risk of becoming extinct in their native range.

MAIN THREATS: Illegal hunting, habitat loss and disease.

ancient ape's descendants ***adapted*** to a wider range of habitats, including rainforests, savannas and grasslands. Some of them even started to live in caves during the hot, dry summer months. Nothing like the shady comfort of a convenient cave to keep a chimp cool!

There are a lot of predators in the chimp world. Chimps are always on the lookout for:

- LEOPARDS, LIONS AND OTHER BIG CATS
- CROCODILES, PYTHONS AND VARIOUS POISONOUS SNAKES
- EAGLES (WHICH ARE KNOWN TO PREY ON BABY CHIMPS)
- HUMANS

ADITYA SINGH/GETTY IMAGES

These clever chimps are using a stick to help get at some tasty termites.
MANOJ SHAH/GETTY IMAGES

Food is also hard to come by, so chimps have a very mixed diet. While their happy hippie cousins the bonobos are vegetarian, chimps are omnivorous. They love fruit more than anything, but along with leaves, bark and seeds, they sometimes chow down on insects, small animals and even, occasionally, other chimps. In response to the constant pressure to find food, chimps have also evolved to be territorial. While they are fun-loving and playful within their own family groups, they can be very aggressive. Chimp groups often fight other troops, and these battles can become violent.

DO CHIMPS REALLY LIKE BANANAS?

They sure do! Bananas are tasty and sweet and nutritious. They are high in ***carbohydrates***, which provide quick energy. Bananas are also loaded with essential vitamins and minerals that help keep the chimps healthy and strong.

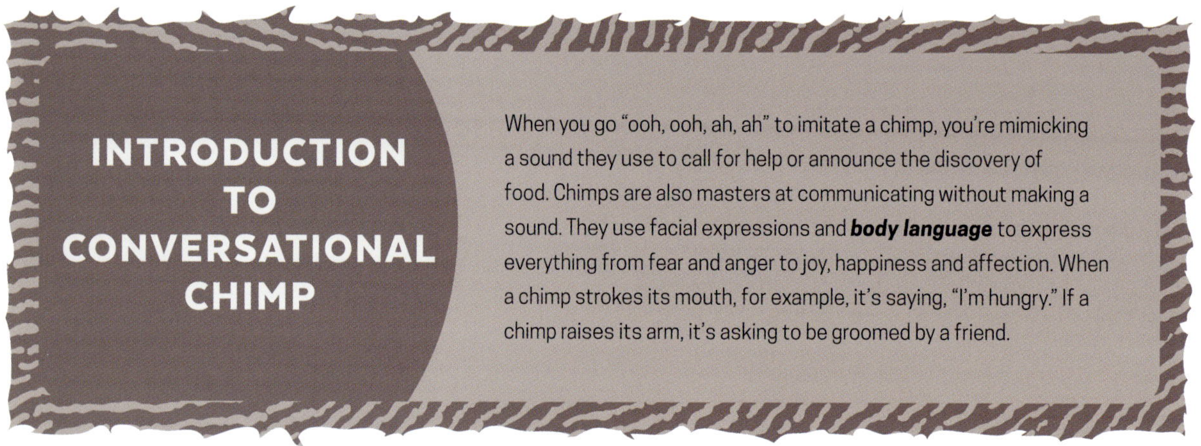

INTRODUCTION TO CONVERSATIONAL CHIMP

When you go "ooh, ooh, ah, ah" to imitate a chimp, you're mimicking a sound they use to call for help or announce the discovery of food. Chimps are also masters at communicating without making a sound. They use facial expressions and ***body language*** to express everything from fear and anger to joy, happiness and affection. When a chimp strokes its mouth, for example, it's saying, "I'm hungry." If a chimp raises its arm, it's asking to be groomed by a friend.

Chimp troops are led by a dominant male.
JOHN BROWN/GETTY IMAGES

BRAINS AND BRAWN

Because of all the dangers they face, chimp communities are close and complex. They live in large groups with families, friends, enemies and all sorts of everyday drama. They love being around other chimps and spend their time playing, grooming one another and working together to find food and build their cozy bedtime nests.

Chimp troops are led by dominant males who assert their authority through displays of power such as charging, drumming on trees and loud vocalizing. The struggle to be leader is ongoing and often becomes violent if one chimp tries to overthrow its rival.

Along with their complex social lives, chimps are great communicators. They use a variety of sounds to express their thoughts and feelings, from hoots, screams and grunts to the famous pant-hoot.

Known for their problem-solving skills, chimps are always looking for ways to make their constant search for food easier. While most animals rely on speed, stealth or brute strength, chimps go a step further, using everyday items such as tools to help them get their hands on their favorite treats. Scientists have watched in fascination as a chimp uses a stick to fish termites out of its nest or scrape honey out of a beehive without getting stung. Now that's what we call smart!

Chimps are great at problem-solving.
GALLO IMAGES/GETTY IMAGES

MORE DARWIN, DARLIN'

ELLIOTT & FRY/WIKIMEDIA COMMONS/PUBLIC DOMAIN

Charles Darwin was controversial in his day. A lot of people didn't believe his idea that all living things were connected through a process he called evolution. He caused an even bigger uproar when he wrote a book called ***The Descent of Man***. Published in 1871, the book took Darwin's theory of evolution and applied it to humans.

Part of the book talked about the things that make humans special. In particular, Darwin believed that our big brains and the way they helped us think, learn and communicate set us apart from all the other animals. But he also said that we shared evolutionary history with other animals. His most controversial idea? He believed that humans and great apes share a common ancestor. He reached this conclusion by doing a detailed comparison of the countless biological similarities between us and our primate relatives.

Many people were angry with Darwin. They twisted his words to make it sound like he'd said that humans descended from monkeys. He calmly accepted the outrage. He understood that his theories challenged religious beliefs about the way the world worked. At the same time, he knew the evidence was overwhelming. He quietly continued his research and, over time, saw that more and more people were starting to understand and accept his ideas about evolution.

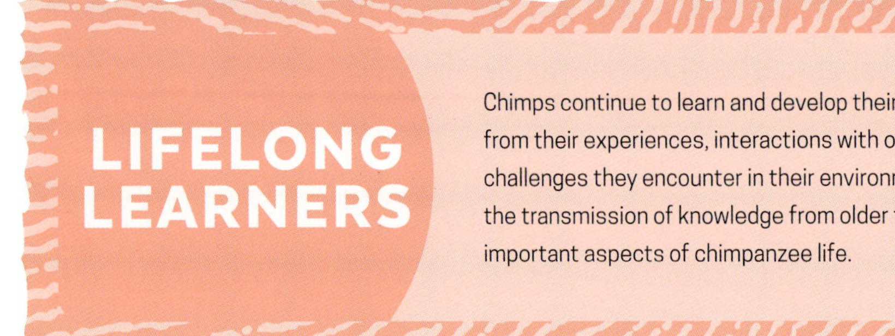

LIFELONG LEARNERS

Chimps continue to learn and develop their whole lives. They learn from their experiences, interactions with other chimps and the challenges they encounter in their environment. Social learning and the transmission of knowledge from older to younger individuals are important aspects of chimpanzee life.

GROWING UP CHIMP

Like bonobos, baby chimps are born with pink faces that darken over time. They are very small at birth and usually weigh less than a pound (half a kilogram)—that's lighter than a can of soda! Newborn chimps, like other great ape babies, have a relatively long period of ***dependency*** on their mothers, becoming independent gradually, over several years.

Chimp children rely on their families to take care of them.
GUENTERGUNI/GETTY IMAGES

The specific time varies based on such factors as their environment, the care provided by their mother and the social dynamics of their group. On average, it takes around five to seven years for a chimpanzee to become fully independent.

During the first year of life, a baby chimp relies completely on its mother for care, nourishment and protection. The mother nurses the infant, carries it on her back and teaches it essential skills for survival. As the chimp grows, it starts to become more independent in terms of movement and exploration. It learns to climb, play and interact with its surroundings. It continues to stay close to its mother but also begins to explore its environment more actively. Eventually the young chimp becomes even more adventurous and starts to interact with other members of the chimpanzee community. It learns important social skills, including how to communicate, cooperate and build relationships with its peers.

Starting at around five years old, chimps become fully independent. They can find their own food, build their own nests and get around their forest world confidently. But even after striking out on their own, chimps maintain strong relationships with their community.

Young chimps are very close to their mothers.
SERGEY URYADNIKOV/SHUTTERSTOCK.COM

PUMPED-UP PRIMATES

Scientists believe that pound for pound, chimps are twice as strong as humans. Of course, when you spend your day swinging through trees, you're bound to get muscular. So don't let their cute faces and smallish bodies fool you. If you arm-wrestle a chimp, you'll lose every time.

CHIMPS IN CRISIS

All four subspecies of chimps are at risk. Scientists classify western, central and eastern chimps as endangered. This means these apes have recently experienced significant population declines and are at a high risk of going extinct in their native habitats. Worse, scientists classify Nigeria-Cameroon chimps as critically endangered. This means their population is so small that they could soon become extinct.

What is causing this decline? Just as with bonobos, the biggest threat to chimps is habitat loss. Their forest homes are shrinking because of logging and farming. As the forests

JANE GOODALL, CHIMP CHAMP

Imagine living in the jungle and studying chimpanzees in their natural habitat. Well, welcome to the world of Jane Goodall, a remarkable scientist who has done more to advance our understanding of chimps than any other person. Goodall was born in England in 1934. As a girl she loved animals and decided to dedicate her life to studying them.

When she was in her twenties, she was finally able to make her dream come true. While visiting a high school friend in Kenya, she decided to reach out to Louis Leakey, a famous ***archaeologist***. The meeting went well, and she soon started working for him. After a while Leakey asked Goodall to start a study of chimps in the wild. Many people didn't think Goodall could do it. She wasn't trained to be a scientist and had no experience living in the rugged mountain rainforests where the chimps lived. It was hard work, especially at first. Whenever Goodall ran into a chimp, it would swing away through the trees as fast as it could.

Eventually the apes started to trust her. Soon she was spending hours every day watching them up close. It was the first time a scientist had studied chimps in their natural habitat...and Goodall's research surprised a lot of people. She saw them using sticks to fish for termites and leaves to soak up water. This was a big discovery, as at the time, scientists believed that only humans used tools. Goodall's findings showed us that animals are much smarter than we once thought.

Goodall also discovered that chimps have complex emotions and distinct personalities. They are playful, loving and even have family squabbles, just like humans. Goodall didn't just observe the chimps—she also worked to protect them. Her research showed how important it is to conserve the chimps' forest homes. Goodall's work with chimpanzees made her famous, and she used her fame to advocate for habitat conservation and animal rights. She founded the Jane Goodall Institute, which continues her research and works to protect chimpanzees in their natural environment.

Jane Goodall discovered that chimps have distinct personalities, just like us.
UNITED ARCHIVES GMBH/ALAMY STOCK PHOTO

get smaller, it's harder and harder for chimps to find food and shelter. The illegal wildlife trade is another problem. Some people capture young chimps to sell as exotic pets. This cruel practice tears young chimps away from their families and prevents many healthy chimps from having babies of their own in the wild. Climate change has a negative effect too. The chimp's forest homes are sensitive to variations in temperature and rainfall, which can make it hard for chimps to find their favorite foods. These changes in the chimps' delicate ecosystem are pushing the already threatened population closer to extinction.

CHIMP SUBSPECIES

While chimps all belong to one species, that group is broken into four smaller subspecies, each with its own traits and behaviors.

WESTERN CHIMPANZEE.

1

Found in West Africa, particularly Guinea, Sierra Leone, Liberia and Ivory Coast, these chimps are very adaptable. Western chimps look slightly different than other chimp subspecies. They often have lighter fur on their bodies, which ranges from light brown to dark black. This color helps them blend in better with their forest homes. Western chimpanzees have specific ways of using tools and vocalizing that are unique to the subspecies. Western chimps also have a taste for different kinds of foods compared to other chimps and tend to eat more fruit than their relatives.

EASTERN CHIMPANZEE.

3

Found in the lush forests of East and Central Africa, including Uganda, Tanzania and the Democratic Republic of the Congo, eastern chimps have some behaviors that make them stand out. They have a special way of cracking nuts, using rocks as tools to open them. This behavior is less common in the other subspecies. Eastern chimps have a diverse diet that includes fruits, leaves and sometimes meat. In fact, these chimps are unique in the way they hunt. While all chimps hunt occasionally, eastern ones do it more frequently, forming small packs to track and kill monkeys, various rodents and even duikers (a kind of small antelope).

(TOP) FIONA ROGERS/GETTY IMAGES; (BOTTOM) ANUP SHAH/GETTY IMAGES

2 CENTRAL CHIMPANZEE.

Found in the heart of Africa, in countries like Cameroon and the Republic of the Congo, these chimps are slender and agile. Central chimps are often found in small, tight-knit groups. One of their unique traits is their nest-building skill. They create beds out of leaves and branches to sleep in, often reusing and improving their nests over time. Like the other subspecies, they have unique vocalizations and eat a varied diet.

4 NIGERIA-CAMEROON CHIMPANZEE.

Found in the lush and diverse rainforests of, you guessed it, the African countries of Nigeria and Cameroon, these chimps are superbly adapted to their forest homes, using their strong arms to swing from tree to tree and gather fruits, leaves and nuts. Unlike other chimps, the Nigeria-Cameroon subspecies comes in a wide range of colors—some are light, and others are dark. Their faces can be hairless or have a full beard of fur. These differences make each chimp unique and beautiful in its own way. They are extremely social, and each community in this subspecies has its own unique gestures, sounds and facial expressions.

(TOP) MARTIN HARVEY/GETTY IMAGES; (BOTTOM) WESTEND61/GETTY IMAGES

Orangutans are masters of the treetops.

FREDER/GETTY IMAGES

4

Orangutans, Our High-Flying Rainforest Friends

ASIA'S ONLY GREAT APE

Orangutans are the gentle giants of the dense tropical rainforests of Indonesia. Known for their remarkable intelligence and impressive physical abilities, they are the only great apes found in Asia.

Orangutans spend almost their entire lives in the dense, interconnected forest canopy, an environment that sets them apart from other great apes. While all apes are good climbers and can move through the forests with ease, orangutans' arboreal abilities are breathtaking. Their powerful arms and long fingers are perfectly adapted for ***brachiation***, allowing them to cover vast distances in their search for food. To help them move through the trees, orangutans have evolved to have truly remarkable reach. Their arms are one and a half times longer than their legs, with a span of 7 feet (2.5 meters) from fingertip to fingertip.

FAST FACTS

NAME ORIGIN: From a Malay word meaning "person of the forest."

SCIENTIFIC NAME: *Pongo* is from the old word originally used to describe any large ape.

SPECIES: Bornean (*Pongo pygmaeus*), Sumatran (*Pongo abelii*) and Tapanuli (*Pongo tapanuliensis*).

SUBSPECIES: Three subspecies of Bornean orangutan—northwest, central and northeast—distinguished by size and the area in which they live.

SIZE: Adult males weigh up to 198 pounds (90 kilograms) and stand about 4.5 feet (1.4 meters) tall. Females weigh up to 82 pounds (37 kilograms) and are slightly shorter than males.

RANGE: Found only in the rainforests of Borneo and northern Sumatra.

HABITAT: They spend their whole lives in trees.

FOOD: Orangutans love fruit but also eat other plants and bark. They will occasionally eat insects and other small animals.

SOCIAL LIFE: Semi-social, with females and young offspring often hanging out together. Males are mostly solitary.

FAMILY LIFE: Females first become mothers between 12 and 15 years old. They have one baby at a time, about every 3 to 5 years.

LIFE SPAN: 35 to 40 years.

POPULATION: 80,000 to 115,000.

CONSERVATION STATUS: Critically endangered. This means they have a very high risk of becoming extinct.

MAIN THREATS: Illegal hunting, habitat loss.

Their unique habitat has shaped their social world. Unlike other great apes, who live in groups, orangutans are largely solitary. Food is hard to find in the treetops, so orangutans have become more independent and self-sufficient. Orangutans also have amazing ***cognitive*** skills. Studies have shown that they can plan ahead and think through problems, and, like chimps, they use tools to help them in their day-to-day lives. Scientists have watched in fascination as an orangutan pokes a termite hill with a stick, then licks off all the juicy bugs. Orangutans are known to make umbrellas out of giant leaves to keep themselves dry in rainstorms, something no other ape (other than humans) is known to do.

MOTHER KNOWS BEST

Even though adult orangutans love to be alone, as youngsters they are very close to their moms. Orangutan moms are true superheroes of the forest. They take care of their babies all by themselves, protecting them from rain and teaching them everything they need to know.

Orangutan kids stay with their moms for the first six to eight years of their lives. During this time they learn all the tricks of the orangutan trade—like swinging from branch to branch and picking the yummiest fruits. Moms are super patient and make sure their kids are ready for the big wide world before they let them explore on their own.

Orangutans do have a social side. When they encounter other orangutans in the canopy, they're not standoffish. They will play and swing together, and they seem to love showing off their climbing skills. But when mealtime rolls around, the orangutans split up and go their separate ways. Of course, being a loner can be a problem when it comes to finding a mate. Swinging through the treetops all by themselves, without social media, how do orangutans meet their future mates? Male orangutans have learned

Orangutans love to show off their swinging skills.

ULET IFANSASTI/GETTY IMAGES

NATURE'S REDHEADS

A life spent in the trees has led to one of the most obvious features of orangutans. Unlike gorillas, chimps and bonobos, which have mostly black fur, orangutans have reddish-brown fur that allows them to blend in better with the surrounding leaves and branches.

TAKE YOUR PICK!

JAMI TARRIS/GETTY IMAGES

Adult male orangutans come in two types, flanged and unflanged. Flanged males have large flat cheek pads and a prominent throat sack. Unflanged males don't, and they're also a little smaller. Why are there two types? Scientists aren't entirely sure, but it probably has something to do with mating. Flanged males use their big puffy throats to make their famously loud long call even louder. They use this call to attract females. And the louder they are, the better their chances of finding a mate.

that it pays to advertise. They let out a loud moan, known as a long call, which rings through the forest and basically says to females, "Hey, baby, I'm looking!" Once they finally do meet up, the male and female orangutan take their time getting to know each other. They swing through the trees side by side, ***forage*** for food together and just hang out, enjoying each other's company.

Sometimes they groom each other, stroking each other's fur and picking bugs off one another. It's the orangutan version of holding hands. But alas, orangutan relationships never last long. Once a female is ready to have a baby, she goes off on her own, builds a cozy nest high up in the trees and settles in to be a full-time mom.

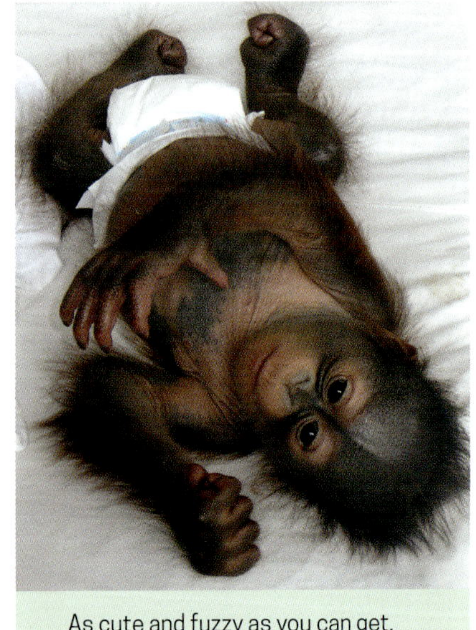

As cute and fuzzy as you can get.
GAVIND/GETTY IMAGES

CHILD OF THE FOREST

Newborn orangutans are tiny, weighing less than a pound (half a kilogram)—just like baby chimps. They are covered in soft, reddish-brown fur. If you're wondering if they're cute and fuzzy, the answer is *yes*!

Each of the three distinct species of orangutan has its own unique traits.

THREE SEPARATE SPECIES

BORNEAN ORANGUTAN. The most numerous of the species, they are found in Borneo, an island in Southeast Asia. They are covered in shaggy, reddish-brown fur and are larger and stronger than other orangutans. Adult males are easy to spot because of their big, protruding cheek pads. Bornean orangutans love to explore their rainforest homes, moving through all sorts of terrain to find their favorite fruits. Lowland rainforests, mountainous areas and even swamps—these are all hangouts of the adventurous Bornean orangutan.

SUMATRAN ORANGUTAN. Rarer than their Borneo cousins, these orangutans are found only in the highlands and mountains of the island of Sumatra. They are smaller and less muscular than the Borneans, and their fur is longer and lighter in color. The best way to recognize a Sumatran orangutan, though, is to check its chin. Both male and females have long, pointy beards. Sumatrans are also more social creatures than Bornean orangutans, and because they live in a safer environment than the Borneans do, with fewer predators, they spend more time on the ground.

TAPANULI ORANGUTAN. This is rarest orangutan species—there are only 800 left in the wild. They are also the most recent species to be identified by scientists. Tapanuli orangutans were discovered in 2017 in the remote South Tapanuli rainforest on the island of Sumatra. How did Tapanulis avoid detection for so long? Partly it's because they look similar to their Sumatran neighbors, so no one realized they were a separate species. But there are small differences. Tapanulis have frizzy hair, smaller heads and flatter faces. However, it wasn't until researchers had the tools to conduct genetic tests that they realized Tapanulis were a species of their own.

(TOP) RAZVAN CIUCA/GETTY IMAGES; (MIDDLE) ANUP SHAH/GETTY IMAGES; (BOTTOM) TIM LAMAN/WIKIMEDIA COMMONS/CC BY 4.0 DEED

BIRUTĖ GALDIKAS: MAKING A DIFFERENCE

"What I have learned from orangutans is that we humans must not turn our backs on our own biological heritage."

—BIRUTĖ GALDIKAS, ***PRIMATOLOGIST***, CONSERVATIONIST

SIMON FRASER UNIVERSITY/WIKIMEDIA COMMONS/CC BY 2.0 DEED

Birutė Galdikas was a little girl with a big dream. As a child growing up in Canada, she was fascinated by books about animals, especially ones about monkeys and apes. Galdikas often imagined what it would be like to live in the wild with these animals and study them up close. As she grew older, her dream never faded. As a young adult, she made a bold decision: she would go to the rainforests of Borneo to study orangutans. With just a backpack, a few supplies and an abundance of courage, Galdikas set off on an adventure that would change her life.

She arrived in Borneo and made her way to the rainforest. Here she encountered orangutans for the first time. She spent hours watching these amazing creatures go about their daily routines. She got to know some of the orangutans personally, and they began to treat her like a friend. Galdikas quickly realized that orangutans were in danger. The rainforests were being cut down to make room for farms and buildings. The orangutans were losing their homes and food sources. Determined to make a difference, she founded Orangutan Foundation International, an organization dedicated to saving orangutans and their habitats.

Since 1986 Galdikas and her team have rescued hundreds of injured and abducted orangutans. The dedicated researchers provide medical care, food and a safe place for these animals to recover. Once the apes are healthy again, they are released back into the wild. Her dedication has paid off. She has helped save countless orangutans and along the way raised awareness about the importance of protecting their rainforest habitat. Galdikas's passion and determination remind us that we all have the power to make a positive impact on the world around us.

Baby orangutans, like other great apes, have a long period of dependency. On average, it takes up to eight years for a baby orangutan to become fully independent. During the first year of life, the baby relies entirely on its mother for care, nourishment and protection. The mother carries the infant everywhere, usually holding it close to her body. As the orangutan grows, it becomes more curious and starts to explore its surroundings. It still stays close to its mother, but it begins to learn basic survival skills such as climbing, foraging and selecting appropriate foods. At around age three, orangutans become more independent and start to spend more time away from their mothers. They become better climbers and figure out better ways to forage for food. This is also the time when young orangutans start to hang out with their peers, learning important social skills.

Eventually they are ready to take care of themselves. They can find food, build nests for sleeping and make their way through the treetops with confidence. But even as they settle into their semi-solitary lives, adult orangutans often keep in contact with the most important ape in their lives: mom.

The bond between an orangutan and its mother never fades.
MANOJ SHAH/GETTY IMAGES

Orangutans are often illegally hunted and sold as pets.
555DIGIT/GETTY IMAGES

GREAT APES FOR SALE

As much as you might like to own your own pet orangutan, it's way more cruel than cool. Unlike dogs, cats and other ***domesticated*** animals, apes are wild creatures that need to live in their native habitats. Unfortunately, there are a lot of people who want to own an ape and will pay big bucks to do so. In fact, the price for an illegally captured gorilla is an unbelievable $400,000.

The United Nations reports that an average of 6,000 great apes are taken from their homes by animal traders every year. Orangutans are by far the most popular, making up 70 percent of all apes captured. Why are they targeted? For one thing, because they are harder to capture and transport, they are more valuable than other great apes. While chimps and bonobos sell for $12,000 to $20,000 on the black market, orangutans fetch $50,000 or more.

Orangutans are also the easiest apes to catch. Because so much of their forest home has been destroyed, they often leave the treetops to search for food. On the ground, it is easier for animal hunters to trap them. Governments around the world have created laws to protect great apes and other wild animals from being taken from their natural habitats. Any animal is at risk, and the more exotic and unusual they are, the greater the risk. The sad truth is that the illegal animal trade is one of the most serious threats faced by endangered species around the world. It's a multibillion-dollar business.

TROUBLE IN THE TREETOPS

Like other great apes, orangutans are facing some big problems. All three species are critically endangered and, in fact, are the most at-risk great apes on the planet. Fifty years ago there were millions of orangutans swinging through the treetops. Now their numbers are much lower. If this trend continues, orangutans might completely disappear from their native habitats in just a few decades. After that the only place you'll be able to see them is in a zoo or nature preserve.

HABITAT LOSS.

Orangutans need a lot of room in which to live and grow. And because they are solitary, each ape has to find its own territory. But as people cut down trees for lumber and farming, and to clear room for palm tree plantations, there is a lot less forest to go around. Without a place to call home, orangutans cannot survive.

CLIMATE CHANGE.

Hotter weather, changes in the amount of rainfall, and disruptions to the usual seasonal patterns—all these things affect plant growth in a forest. This makes it harder for orangutans to find their favorite foods, so they must expand their territories even more, in search of the fruits, leaves and branches they are used to eating.

ILLEGAL ANIMAL TRADE. The United Nations reports that more than 4,200 orangutans are captured every year. These poor animals are shipped around the world to become pets, circus performers or subjects for experiments in research labs.

Given all the problems orangutans face, a lot of people are trying to help them. Organizations like Orangutan Foundation International are working to protect their habitats, stop animal trading and raise awareness about these amazing acrobatic apes.

Clear-cutting forests is making life harder for orangutans.
MANGIWAU/GETTY IMAGES

STAY CALM, AVOID PALM

Palm oil is good news for our health. While other types of cooking oil can contribute to heart disease, stroke, obesity and other medical problems, palm oil is cholesterol-free (that's good) and has a lot of vitamins and other ***nutrients*** that help us stay healthy. Palm oil has become very popular. More than half of all the cookies, ice cream and other snack foods you like to eat are made with palm oil. Even toothpaste, lipstick and shampoo use palm oil.

Another great thing is that palm oil comes from the fruit of palm trees (duh), which produce a lot of oil but use only a small amount of land. The trees grow best in tropical climates, and palm farming is providing jobs and money for people in poorer

Rainforests are being cut down to make room for palm oil plantations.
KAMPEE PATISENA/GETTY IMAGES

When we make good choices, it can help build a better future for our orangutan friends.
GUENTERGUNI/GETTY IMAGES

parts of the world, like Southeast Asia and West Africa.

The bad news is that palm farming has exploded over the past 20 years. Palm plantations are shooting up everywhere. And the same kind of weather and geographic conditions that make good habitat for great apes are also good for palm trees. As a result, a lot of rainforest has been cut down to make room for palm plantations, leaving the great apes with no place to go. Orangutans have been the hardest hit. Over the past 10 years, 80 percent of orangutan habitat as been altered or destroyed to make room for palm trees.

You can make a difference. Encourage your family and friends to buy only products made with ***sustainable*** palm oil. This means the trees were grown on a plantation that respects existing animal habitat and treats its workers fairly. Likewise, there are lots of healthy alternative oils. Olive, coconut, canola—all these oils have health benefits and come from farms that don't disrupt the orangutans' way of life.

Gorillas are the gentle giants of the jungle.
HENLLY TATA/GETTY IMAGES

5

Gorillas, Gentle Giants of the Jungle

BIGGER IS BETTER

The heart of Africa's lush and mysterious rainforests is home to the largest great apes, the gorillas. With their sheer size, gentle demeanor and intricate social lives, gorillas are one of nature's most fascinating creatures.

Gorillas are big. A dominant male is as big as a man and weighs more than twice as much. They're also very, very strong. A full-grown male gorilla is as strong as 10 humans! Because of their size, gorillas have developed a unique way of getting around when they are not hanging out in trees. They are ***quadrupeds***, using their enormous arms like an extra pair of legs that supports their great weight.

Despite their ferocious appearance, gorillas are shy and gentle. One would attack you if threatened, but even then, the gorilla would put

FAST FACTS

NAME ORIGIN: An ancient Greek name for a mythic tribe of hairy warriors.

SCIENTIFIC NAME: *Gorilla*

SPECIES: Two, eastern (*Gorilla gorilla*) and western (*Gorilla beringei*), each having two subspecies.

SIZE: Adult males can be up to 6 feet (1.8 meters) tall and weigh a whopping 500 pounds (225 kilograms). Girlillas top out at 4.5 feet (1.5 meters) and weigh up to 200 pounds (115 kilograms).

RANGE: Cameroon, Central African Republic, Gabon, Republic of the Congo, Democratic Republic of the Congo, Equatorial Guinea.

HABITAT: Lowland and mountain tropical rainforests. Gorillas are occasionally arboreal but spend most of their lives on land.

FOOD: Mostly herbivores, eating leaves, shoots, stems, flowers and bark. Gorillas will occasionally eat larvae, snails, ants and small insects.

SOCIAL LIFE: Usually live in small family groups with up to 10 members, headed by a dominant male called a silverback (for the patch of silvery hair that grows across his back).

FAMILY LIFE: Females start having babies around age 10 and have one baby at a time every four to six years.

LIFE SPAN: 35 to 40 years in the wild, although 50-year-old gorillas are not uncommon.

POPULATION: 350,000 western, 5,000 eastern.

CONSERVATION STATUS: Three of the four subspecies are critically endangered. This means they have a very high risk of becoming extinct.

MAIN THREATS: Illegal hunting, deforestation, disease.

on a show of strength first, roaring and pretending to charge. As long as you backed down and let the gorilla know you were friendly, it would back down.

AN EVOLUTIONARY TALE

How did gorillas evolve into the gentle giants of the jungle? The earliest gorilla-like ancestors appeared eight million years ago. These animals were likely smaller than modern gorillas and lived in both forests and grasslands. Over time gorillas adapted to become forest dwellers. They became larger and more muscular, which allowed them to bulldoze their way through the thick forest underbrush. They also developed powerful jaws and long, sharp teeth for chewing through branches, thick leaves and bark.

ONE BIG, HAPPY GORILLA FAMILY

Just as their bodies evolved over time, so did their social habits. They developed close family groups that live cooperatively. Living in these troops helps keep gorillas safe. Gorillas form

Gorillas form strong bonds within their family groups.
IBRAHIM SUHA DERBENT/GETTY IMAGES

strong bonds within their troops. They groom each other, picking dirt and bugs off one another, which helps them bond and keep clean. It's like a gorilla version of cuddling.

A typical gorilla troop consists of multiple individuals led by a dominant male, known as the silverback. The silverback is usually the biggest and strongest of the group and makes important decisions to ensure the troop's well-being. The silverback is not just the leader but also the protector. He guards the troop against danger, whether it's from other animals or human activities. He's the one who decides where the group will go to look for food and when it's time to rest. The rest of the gorillas in the troop look to the silverback for guidance and protection.

Gorilla troops also include several females, their offspring and sometimes a few other males. The females

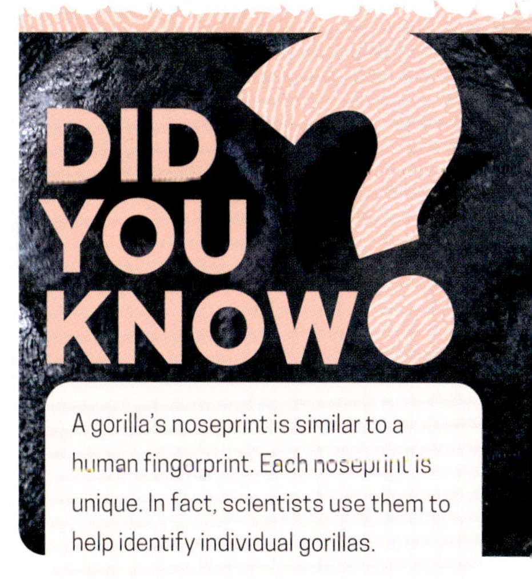

A gorilla's noseprint is similar to a human fingerprint. Each noseprint is unique. In fact, scientists use them to help identify individual gorillas.

ONE GORILLA, TWO GORILLA, THREE GORILLA, FOUR...

Each gorilla species has two subspecies.

EASTERN MOUNTAIN GORILLA. Mountain gorillas are found in the highlands of Uganda, Rwanda and the Democratic Republic of the Congo. They have wide chests and muscular arms and are known for the dark, thick fur that protects them from the chilly mountain weather.

EASTERN LOWLAND GORILLA. The largest members of the primate family, these gorillas are stockier than mountain gorillas, with large hands and shorter muzzles. They are found in the lowland forests of the eastern part of the Democratic Republic of the Congo.

WESTERN LOWLAND GORILLA. The most populous on Earth, there are 100,000 of them spread out across Gabon, Central African Republic, Cameroon, Angola, Equatorial Guinea and the Congo basin. They are slightly smaller than other subspecies and have brown-gray coats with light-red chests. Fully grown dominant males have a distinctive silver patch on their backs that runs from their shoulders right down to their rumps.

WESTERN CROSS RIVER GORILLA. This subspecies is limited to a small patch of mountain forest on the border of Cameroon and Nigeria. There are fewer than 300 of these gorillas left in the wild. They are smaller and slighter than other gorillas and have light-black hair.

(TOP TO BOTTOM) MARTIN HARVEY/GETTY IMAGES; GUENTERGUNI/GETTY IMAGES; FIONA ROGERS/GETTY IMAGES; ROBERT OCH/GETTY IMAGES

TWO SEPARATE SPECIES

There are two gorilla species, western and eastern. They live in different parts of ***equatorial Africa***, and there are two subspecies of both. As their names imply, the two species evolved in different parts of Africa. Although they look similar, the best way to tell eastern and western gorillas apart is by looking at their fur. Eastern gorillas tend to have darker and thicker fur because they live in cooler mountain areas. Western gorillas, on the other hand, have lighter-colored, thinner fur because they live in warmer forests.

While all gorillas are social animals that like to hang out in groups, the eastern variety likes to keep things smaller. Five to 10 members per troop is the norm for them. Western gorilla families are often bigger, with around 10 to 30 members. Both eastern and western gorillas love to eat plants, but they have slightly different tastes. Eastern gorillas munch on leaves, stems and fruits. They're healthy eaters who love their veggies! Western gorillas also eat plants, but they're a bit more flexible. Their diet might include occasional treats like termites and ants.

are often related to one another and can be the silverback's daughters, sisters or nieces. Young males, called blackbacks, stay with the troop until they are old enough to start their own groups.

GROWING UP IN THE JUNGLE

Even though adult gorillas are the biggest of the great apes, baby gorillas are the same size as newborn bonobos, chimps and orangutans. And they're much smaller than newborn humans. You probably weighed between 6 to 8 pounds (2.7 to 3.6 kilograms) when you came out of your mom's tummy. A little one-pound (half-kilogram) gorilla would look tiny beside you.

Just like other apes, including us humans, baby gorillas depend on their moms for almost everything. Food, love, protection—mama gorilla provides it all. As the baby gorilla

Fatou, a western lowland gorilla living in Zoo Berlin, celebrated her 66th birthday on April 13, 2023. She's the oldest living gorilla on record.

Baby gorillas depend on their moms for almost everything.
JANE RIX/SHUTTERSTOCK.COM

grows, it becomes more active and starts to explore its environment. It begins to eat foods in addition to nursing. The mother gorilla helps the infant learn what foods are safe and nutritious. At around three years old, young gorillas become more involved in the group's activities, such as playing with other young gorillas and observing adult behaviors.

As the gorillas continue to grow and learn, they become increasingly self-reliant. While they may not be completely independent until around eight to ten years old, the process of acquiring the skills and knowledge needed for survival continues throughout their early years. Gorilla groups provide a supportive environment in

which young gorillas learn and grow. The silverback plays a protective and guiding role, and young gorillas benefit from interactions with other members of the group. Even after becoming independent, gorillas stay part of their family group.

CONSERVATION: NOT ALL BAD NEWS

Like other great apes, gorillas face serious challenges to their survival as a species. Both subspecies of western

Silverbacks lead and protect a gorilla troop.
KEVIN SCHAFER/GETTY IMAGES

gorillas are critically endangered and are at risk of soon disappearing in the wild. The lowland subspecies of the eastern gorilla is also critically endangered. However, the mountain subspecies is slowly making a comeback. In 2018, conservationists upgraded that subspecies from critically endangered to endangered. That means that even though these apes are facing a lot of challenges, things are slowly improving.

As to why gorillas are all at risk, it's because of habitat loss caused by deforestation, ***poaching*** and climate change. But thanks to organizations like the Dian Fossey Gorilla Fund, gorillas and their homelands are being protected. Conservation efforts have helped the eastern mountain gorilla, proving that if we all work together, we can make a difference.

Gorillas face serious challenges to their survival as a species.
LEONARDOSPENCER/GETTY IMAGES

DIAN FOSSEY: GORILLA GUARDIAN

Dian Fossey was a hero of science, who dedicated her life to studying—and protecting—the majestic mountain gorilla. Born in California in 1932, Fossey had a tough time as a kid. Her parents divorced when she was very young, and she lost contact with her father. She was also extremely shy. Maybe this is why she became interested in animals, and by the time she finished school she had decided she wanted to become a veterinarian.

On a holiday in Rwanda, Fossey saw wild gorillas for the first time at a wildlife reserve. Something about these majestic apes struck her. She knew then and there that she wanted to spend her life studying and helping them. A little later Fossey met Louis Leakey. He was a famous scientist who had encouraged both Birutė Galdikas and Jane Goodall to study apes in the wild. Impressed with Fossey's passion for gorillas, Leaky encouraged her to do field research of her own.

For the next 18 years Fossey lived in the Rwandan mountains, spending her time with gorillas, studying and recording their behavior. She became close to the apes and had special names for each of them. Fossey faced many challenges during her time in Rwanda. She had to overcome language barriers, adapt to harsh living conditions and learn how to protect gorillas from the poachers who hunted them for food. Along the way, she started the Karisoke Research Center, a place where people could study gorillas in the animals' natural habitat.

Eventually she wrote a book about her experiences. ***Gorillas in the Mist*** was a bestseller around the world and got people thinking about how they could help mountain gorillas. Not everybody appreciated Fossey's hard work. In 1986 she was found dead in her remote mountain cabin. Someone had attacked and killed her. No one knows for sure who did it or why, although many people think it was because of her work to protect gorillas from being illegally hunted.

But her legacy carries on. The research center she founded continues to study these wonderful primates, and the Dian Fossey Gorilla Fund is helping protect habitats and has saved countless gorillas from being hunted.

LIAM WHITE/ALAMY STOCK PHOTO

CAN APES REALLY TALK?

Koko the gorilla was fluent in American Sign Language (ASL). Kanzi the bonobo could understand more than 1,400 words. But were they really talking? There's no question that great apes are amazing communicators. In the wild, they have a bunch of different ways to tell their friends what they are thinking and feeling. Gestures, facial expressions, songs, grunts and other vocalizations and good old body language—great apes use all these techniques to get their messages across.

But when it comes to using sign language or, as Kanzi did, lexigrams, does it represent creative, spontaneous use of language? Or are these examples of apes giving automatic answers in hopes of a reward? Some people argued that Koko and Kanzi were going through the motions, putting together strings of words without any real understanding of what they meant. There was also the issue of the human translators. It was often up to the animals' trainers to explain the meaning of what Koko and Kanzi had communicated. But who's to say their explanations were accurate?

On the other hand, there are people who are convinced that the apes were more than copycats: they sometimes combined words in unique ways that suggested a level of understanding that went beyond mimicry. In the end, nobody is really sure. What we do know is that both Koko and Kanzi showed a tremendous ability to express themselves. Their stories have encouraged researchers to look much deeper into the minds of animals to try to better understand how they think and feel.

PARLEZ-VOUS GORILLA?

Just like humans, gorillas have their own way of talking to each other. They use a combination of sounds, facial expressions and body postures to communicate. For example, they might make a low grumbling sound to show they're happy or stand up on two legs to assert their dominance. When it's time to start a family, female gorillas are very choosy. They prefer silverbacks who are strong and kind. The silverback's skill as a protector and leader is crucial to attracting females.

DO GORILLAS BEAT THEIR CHESTS?

You've seen it in cartoons and hokey jungle movies. An angry gorilla, standing upright, beats its chest as it hoots and grunts. In fact, it's something dominant male gorillas actually do. It's not just to show who's boss. The tone and pitch of the sound, which can be heard up to a mile way, tell potential rivals how big and strong the chest beater is. It's both a warning and a challenge. It's the gorilla way of saying, "If you think you can take me, step on up. Otherwise, stay out of my way!"

HOW DO GORILLAS SAY HELLO?

Just like humans, gorillas like to say hello when they meet one of their friends. Often they'll give each other a hug or maybe touch noses, which is the gorilla equivalent of a handshake.

Koko "speaks" to her caregiver using ASL.
SAN FRANCISCO CHRONICLE/HEARST NEWSPAPERS/GETTY IMAGES

KOKO, THE GORILLA SUPERSTAR

Koko, a western lowland gorilla, was a remarkable ape with an extraordinary ability to communicate using ASL. Throughout her long life, she captured the public's imagination. Koko was born at the San Francisco Zoo in 1971. She was a curious young ape with a keen mind. When she was just a few months old, she moved to a research facility where she was surrounded by human caretakers. It was part of an interesting experiment. Scientists wanted to find out if they could teach a gorilla to use human language to communicate its thoughts and feelings.

The scientists decided to try ASL, a system of hand signals used by people with hearing impairment. They hoped this might overcome the fact that gorillas don't have the ability to use spoken language and maybe be the key to bridging the interspecies language barrier. Koko picked up the basics of sign language quickly. By the time she was a teenager, she knew more than 1,000 words and could even combine groups of words to make sentences.

Many of the words Koko used were directly related to her everyday life. She could ask for her favorite foods. She could say what she liked and didn't like. She could describe basic features of people in her life. She also seemed to understand emotions. She could use words like ***happy***, ***sad*** and ***angry*** to express how she was feeling. The experiment surprised a lot of people. Many scientists believed that only humans were capable of using language. Koko proved them wrong.

Koko also had a tender side. She was particularly close to the kittens who lived with her in the research center. Photos of Koko cradling and kissing her beloved pets warmed the hearts of people around the world. Koko showed us that even wild animals like gorillas have feelings just like we do.

Every great ape species is facing serious challenges.
USO/GETTY IMAGES

6

What's Next:

Great Apes on the Edge and How You Can Help Them

Here is the sad truth: great apes are in trouble. Every single species and subspecies is endangered or worse. This means that one day soon, if we don't take action to help them, there will be no more great apes in the forests of Africa and Asia.

As we've already seen, the biggest problem is habitat loss. Great apes are connected to the natural world that surrounds them. But that world is shrinking. Because of deforestation, more than half of great ape habitats have disappeared over the last 100 years. Some areas have been hit harder than others. The orangutans' rainforest home is 90 percent smaller now then it was just 25 years ago.

As ape populations decline, human ones are growing. More people means a greater need for roads, power lines and railway tracks. These things are often cut through forests and jungles, disrupting the way of life great apes have enjoyed for thousands of years. And with more apes and humans living close to one another, there is a higher risk of diseases spreading across species. Polio, pneumonia, measles and even the deadly Ebola virus have all moved from humans to apes.

POPULATION PROBLEMS

Each year thousands of great apes are taken from their forest homes. Some are sold in the illegal pet trade. Thousands more are killed for food. Complicating the issue is the fact that great apes reproduce very slowly. A female has one baby every four to six years. In her entire life, she will probably have no more than four babies. Female dogs, by comparison, can easily have 70 puppies or more! While the slow birth rate gives moms a chance to care and nurture their kids, it also makes it harder for the ape population to grow.

As humans spread out, we are ruining great ape habitats.
DIMAS ARDIAN/GETTY IMAGES

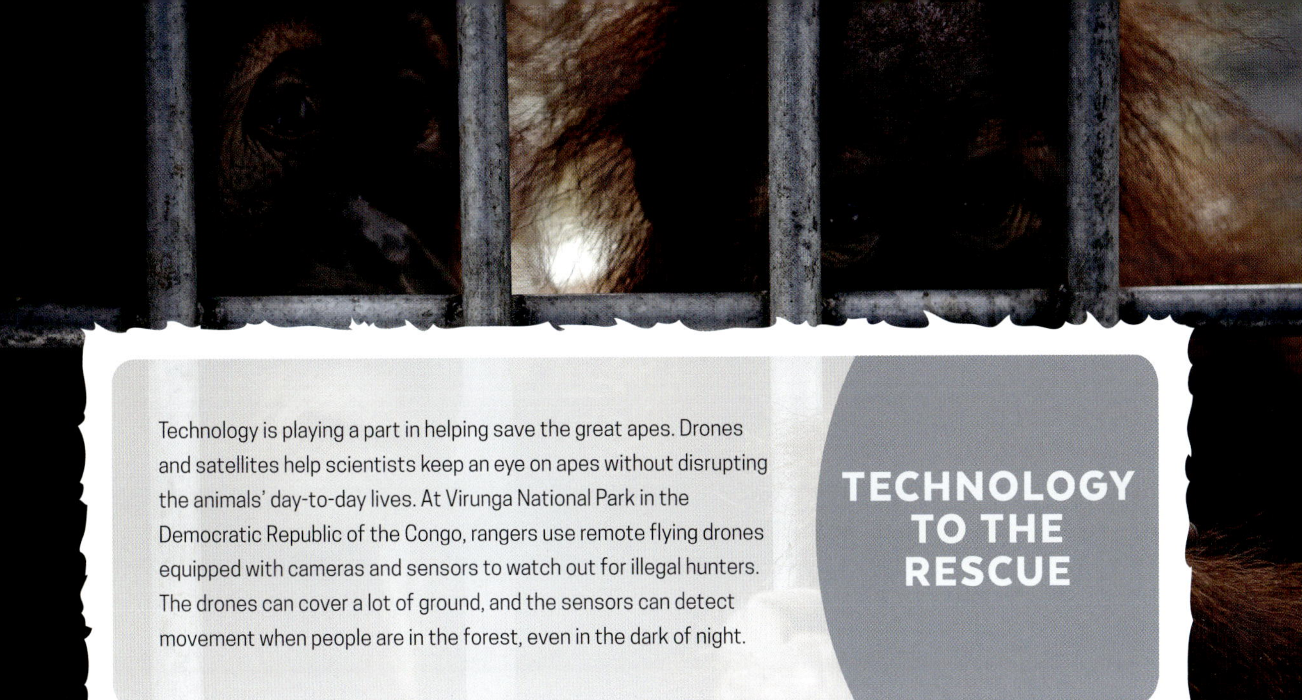

TECHNOLOGY TO THE RESCUE

Technology is playing a part in helping save the great apes. Drones and satellites help scientists keep an eye on apes without disrupting the animals' day-to-day lives. At Virunga National Park in the Democratic Republic of the Congo, rangers use remote flying drones equipped with cameras and sensors to watch out for illegal hunters. The drones can cover a lot of ground, and the sensors can detect movement when people are in the forest, even in the dark of night.

The fight against illegal trapping continues.
ULET IFANSASTI/GETTY IMAGES

ALL FOR ONE

There is hope for great apes, though. Governments, conservation organizations, communities, scientists and ordinary people—if we all pull together, we can restore ape habitats and protect the health and safety of our wild cousins.

PROTECTED AREAS

One thing that's making a difference is the creation of protected areas where apes can live in safety. These protected areas are like special playgrounds for great apes, where they can be free to live their lives without the threat of losing their homes.

The United Nations Educational, Scientific and Cultural Organization (UNESCO) supports a program that designates biosphere reserves and world heritage sites, which are large areas, sometimes spanning more than one country, in which ape habitats are protected. The Gunung

Protected areas are safe spaces for great apes. Reserves like this one are helping protect orangutan homes.

(MAIN) ALVAROBUENO/GETTY IMAGES; (INSET) BAS VERMOLEN/GETTY IMAGES

Leuser Biosphere Reserve, for example, a national park that is part of the Tropical Rainforest Heritage of Sumatra in Indonesia, protects the traditional home of the Sumatran orangutan. UNESCO also created the Gombe Masito Ugalls Biosphere Reserve in Tanzania, in the same area where Jane Goodall started

Everyone can make a difference—even you!
JENNY EVANS/GETTY IMAGES

BE PART OF THE SOLUTION

Small things can make a big difference. Here are a few simple steps you can take to help great apes and also support conservation efforts in your own hometown.

Make ape-friendly choices. Encourage your family to choose products made with sustainable palm oil or palm oil alternatives. Unsustainable palm oil production destroys orangutan habitat.

- You don't have to go to Africa or Asia to show you care for great apes. You can start at home by supporting local conservation projects.
- Reduce and recycle your garbage. Join community cleanup events to help keep the local environment safe and healthy.
- Plant trees and gardens! Just as trees are essential to great ape habitats, they're also an important part of local wildlife habitats. Join in tree-planting activities or create a small garden to support habitats in your area.
- Use eco-friendly products like reusable water bottles, lunch containers and shopping bags.

THE MORE THE MERRIER

- The name for a general group of apes is a ***shrewdness***.
- A group of bonobos is called a ***troop***.
- A group of chimps is a ***whoop*** or ***troop***.
- A gathering of orangutans (as rare as that is) is a ***congress*** or ***buffoonery***.
- The name for a group of gorillas is ***band*** or ***troop***.

chimp research more than 60 years ago. At the time of this writing, there are 36 UNESCO-designated sites in 23 countries to help protect great apes and their habitats.

Education is an important tool. When people better understand great apes, their habitats and the challenges they face, they are better equipped to help make positive changes.

EASTERN MOUNTAIN GORILLAS: A SUCCESS STORY

Forty years ago there were only 250 eastern mountain gorillas left in the wild. Extinction seemed unstoppable. Conservation efforts were stalled. The Congo basin where these apes lived was the scene of a brutal civil war. Anyone who tried to help the gorillas was putting their life at risk. Despite the dangers, conservationists, scientists, local communities and governments joined together to try to protect these majestic apes. They created a reserve in Virunga National Park where gorillas and other animals can live with some measure of protection.

And as the park recovers from years of war, the World Wildlife Fund, an international conservation organization, has stepped in, raising money for reforestation and hiring special park rangers to protect the mountain gorillas from poachers.

The results have been amazing. Ten years ago, there were fewer than 500 gorillas in Virunga National Park. Now there are over 600 of them, and eastern mountain gorillas have moved from being classified as critically endangered to endangered. There's still a long way to go, but the success in Virunga is proof that, working together, we can make a difference.

Today there are more than 600 gorillas in Virunga National Park.

(TOP) KONRAD WOTHE/GETTY IMAGES; (BOTTOM) BRENT STIRTON/GETTY IMAGES

GENTLE JAMBO

Jambo was a silverback gorilla whose actions touched the hearts of people around the globe. It's a story of how intelligent, caring and compassionate great apes really are. It all began one hot summer day back in 1986 at the Jersey Zoo in the United Kingdom. Five-year-old Levan Merritt was trying to get a better view of the gorilla enclosure. As he leaned over the edge, he suddenly slipped and dropped about 15 feet (4.5 meters) to the ground below.

A crowd of people stood breathless as an enormous silverback gorilla approached the boy, who'd been knocked unconscious by the fall. Some people thought that Jambo was going to attack Levan. The boy was, after all, intruding on the dominant male's territory. Instead Jambo carefully took a place by Levan's side, protecting him from the other apes. He gently stroked the boy's back and waited for help to come.

Just as Levan came to and started to cry, a zookeeper and paramedic jumped into the enclosure. Jambo wisely moved away, giving the workers a chance to help the injured boy. Jambo gathered his family and moved them all to a hut in the corner of the pen, where they stayed until Levan had been safely removed.

JAMBO GOES VIRAL

By chance, one of the zoo visitors had a video camera and filmed the entire scene. Soon people around the globe were watching Jambo's gentle heroics. He was a viral sensation (before social media), showing the world that even a massive wild animal like him was capable of tenderness and compassion.

Jambo became a famous symbol of the bond between humans and animals. Thousands of people visited Durrell Wildlife Park every year just to see him. Levan Merritt spent six weeks in hospital recovering from a concussion and broken arm. But he never forgot Jambo's kindness and often visited the zoo to show his appreciation.

In 1992, after Jambo died of old age, Merritt took part in a special ceremony at the wildlife park honoring this amazing ape. He cut the ribbon, officially unveiling a statue of Jambo, and said a few words thanking the sensitive silverback who reminded the world that all of us—bonobos, chimpanzees, orangutans, gorillas and humans—are part of the same big, wonderful family of great apes.

BE AN APE AMBASSADOR

Remember Molly, the bonobo I encountered in a small, sad-looking zoo in the mountains of British Columbia? As a kid, I knew I wanted to help Molly, but I really didn't know how.

It's easier today for people to make a difference. You don't have to fly to Asia or Africa and live among the apes. You don't have to spend millions (or even tens) of dollars. You don't have to go on TV and make speeches. All you have to do is make a simple decision to take action.

Maybe that's why you picked up this book in the first place. You're already interested in helping great apes, and you want to learn more about them. Keep going! There are lots of other books, websites, TV shows and videos out there that can tell you the latest facts and figures from the great ape world.

"I have learned you are never too small to make a difference."
—Greta Thunberg, climate activist

STEFANO GUIDI/GETTY IMAGES

EDUCATING YOURSELF IS A GREAT WAY TO START

The more you know about great apes, the better able you'll be to make a difference. What exactly can you do? You can start by spreading the word. Tell your friends, family and classmates about how great these great apes are and how badly they need our help. You can also find ways to support your favorite conservation organizations. Visit their websites or follow them on social media. The more followers an organization has, the more influential it can be. You can also tell your friends about these organizations and promote them on your social media. And while you're at it, why not put on a fundraiser at your school? Bake sales, art exhibits, fun runs—these are all easy and enjoyable ways to raise money *and* awareness. The future of great apes is in our hands. By working together, we can create a world where these incredible animals continue to swing through trees, communicate with each other and thrive in their natural homes. Whether it's through conservation efforts, technological advancements or simply making mindful choices, we have the power to shape a future where great apes—and all species—share a harmonious world. So let's join forces and do what we can to help protect our amazing great ape family.

GET CONNECTED

Here are some of the amazing organizations around the world that are helping great apes. Check out their websites for facts, fun activities and videos, and ideas on how to be a better ape ambassador and conservation champion.

CONSERVATION

WORLD WILDLIFE FUND. A global conservation organization helping protect endangered species and their habitats around the world. Their work includes protecting elephants from the illegal ivory trade, restoring giant panda habitat and encouraging countries around the world to help save belugas and other at-risk whale species. They've also led numerous programs to protect great ape habitats.

RANGER RICK. A conservation magazine just for kids. Follow the adventures of Ranger Rick the Raccoon and his friends as they help endangered species around the world. The website offers games, videos, stories, crafts and previews of the *Ranger Rick* and *Ranger Rick Jr.* books and magazines.

BONOBOS

BONOBO CONSERVATION INITIATIVE. The only international organization dedicated to protecting bonobos. The Bonobo Conservation Initiative has been around since 1998 and so far has managed to protect and restore 9 million acres (3.6 million hectares) of bonobo habitat. The website has cool features like audio recordings of bonobos vocalizations, stories and even songs.

CHIMPS

JANE GOODALL INSTITUTE. Goodall started the organization in 1977 to do research on chimps and help protect them from habitat destruction and the illegal pet trade. Today the institute uses high-tech equipment to monitor chimps in the wild. Goodall's team also rescues wild chimps from pet dealers and rehabilitates the apes for release in the wild. So far the organization has protected thousands of acres of chimp habitat, creating a safe home for 5,000 chimps and gorillas. The organization also runs Roots & Shoots, a special group that helps kids lead conservation projects in their own communities.

> “Every individual has a role to play. Every individual makes a difference.”
>
> —JANE GOODALL

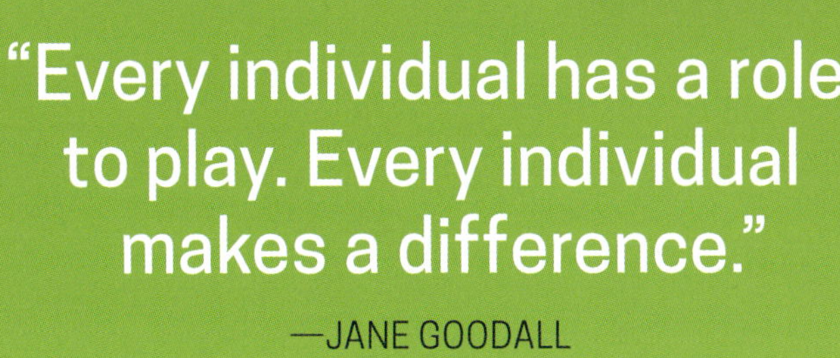

ORANGUTANS

ORANGUTAN FOUNDATION INTERNATIONAL. Started by Birutė Galdikas in 1986, the organization works to keep wild orangutans in their rainforest habitat. The organization also supports research and education programs that raise awareness about orangutan conservation. One of the coolest things about Orangutan Foundation International is its Eco Tours program. Imagine joining Galdikas in the mountains of Borneo, helping with her conservation work. Who knows, you might even meet an orangutan or two along the way.

GORILLAS

DIAN FOSSEY GORILLA FUND. Founded by Fossey in 1967, this organization monitors and protects gorillas in their natural habitats. The fund also runs anti-poaching teams that watch over the gorillas and protect them from illegal hunting. The organization’s efforts have helped move the eastern mountain gorilla off the critically endangered list.

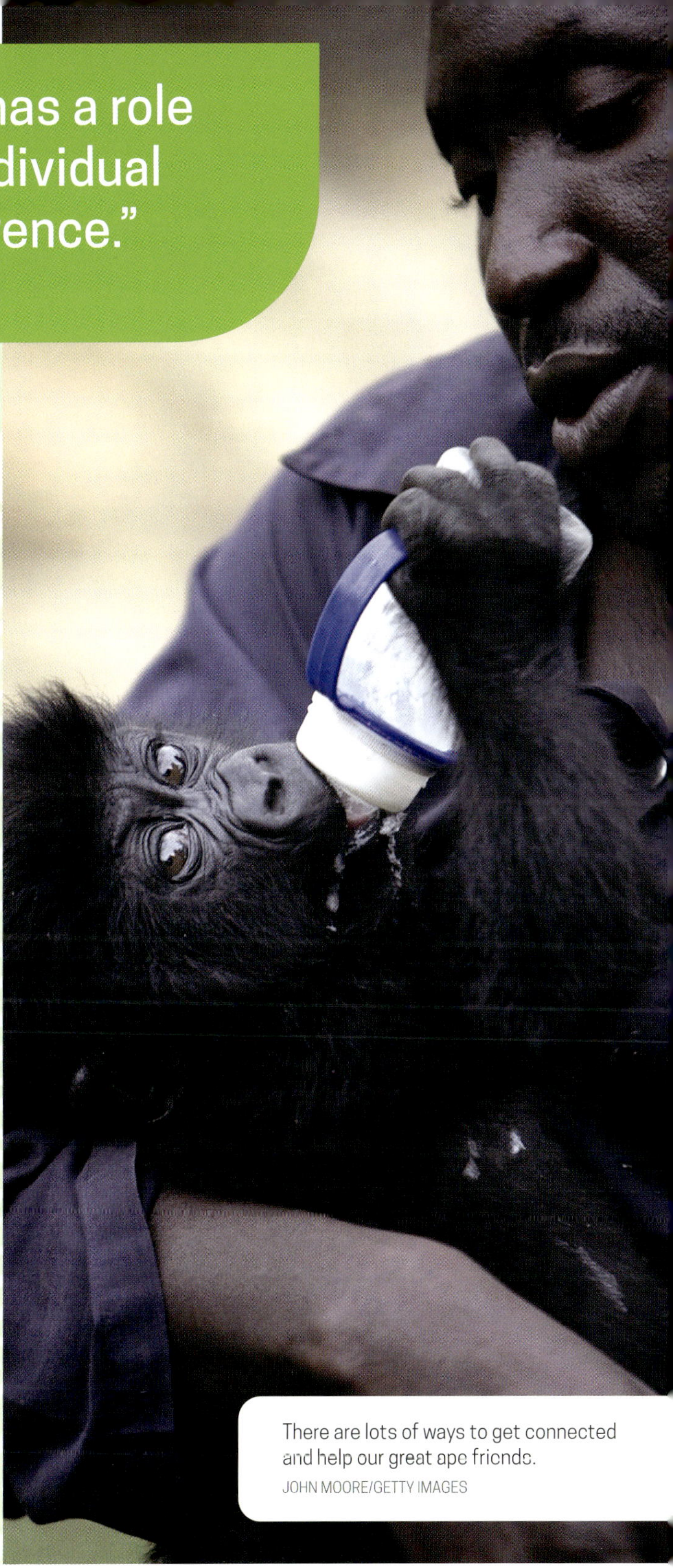

There are lots of ways to get connected and help our great ape friends.
JOHN MOORE/GETTY IMAGES

Glossary

adapted—changed in response to changes in the environment to become suited to it. Adaptation is one of the key concepts of evolution.

Age of Dinosaurs—a period from about 245 to 66 million years ago, known as the Mesozoic era

arboreal—living in or associated with trees

archaeologist—a scientist who examines very old things, like buildings, tools and bones, to better understand how people lived in the past

BCE—short for Before the Common Era

body language—the gestures, movements and facial expressions that a person or animal uses to communicate with others. More than half of the things we communicate are expressed without words.

brachiation—progression by swinging from hold to hold using the arms. It's how most apes get around.

carbohydrates—the body's main source of energy, consisting mainly of sugars (such as what's found in fruit, milk and candy) and starches (found in grains, bread and pasta). Apes love carbs!

cells—the smallest units of life in an organism that are capable of existing by themselves. Every living thing is made up of cells, each of which has a specific function, from helping you see to helping you breathe.

cognitive—relating to mental processes such as thinking, remembering and understanding

deforestation—the process of cutting down trees and also the state of a forest having been cleared

dependency—the state of relying on and needing someone or something else to survive

DNA—short for deoxyribonucleic acid (dee-OCK-see-ri-bo new-KLEE-ik), the genetic material in cells that carries the information that determines how we look and function. The tip of your baby finger is five million times wider than a strand of DNA.

domesticated—evolved over thousands of years from a wild or natural state to live or work with humans

ecosystem—the community of all the living and nonliving things in a specific area. Everything in an ecosystem is connected, so damaging one aspect of it may cause damage to its other parts.

empathy—the ability to understand and be sensitive to another's feelings or situation

equatorial Africa—those parts of Africa that run along the equator

evolution—the process of changing over time, which leads to new species or populations as living things adapt, over generations, to changes in their environments

extinct—no longer existing

forest canopy—the dense layer of intertwined leaves and branches at the top of a forest

forage—to search for food, especially fruit, leaves, nuts, berries and plants

genes—the smaller parts inside every string of DNA. Each gene carries special chemical information that tells your body how to create specific traits, like eye color, foot size and even how much you sweat. Each strand of human DNA contains 20,000 to 25,000 genes.

germinate—to grow from a seed into a young plant

grasslands—areas that are generally flat and covered in various kinds of grasses, with few or no trees

habitat—the natural place where a plant, animal or other organism normally lives

herbivores—animals that eat plants

hominoids—members of the superfamily of animals with human-like traits, including living animals, like humans, great apes and monkeys, and extinct species like Neanderthals

hybrid—a mix of two different species or subspecies

matriarchs—females in charge of families or groups

nutrients—vitamins and minerals that help keep our bodies strong and healthy

omnivorous—feeding on both plants and animals

opposable thumbs—thumbs that can easily touch the tips of the other fingers. This allows an animal to grasp and hold on to things more effectively. All great apes (and other animals like some frogs, koalas and pandas) have them.

poaching—illegally catching or killing an animal

primates—animals in the group of mammals that includes lemurs, monkeys, apes…and humans

primatologist—a scientist who studies primates like monkeys and apes

quadrupeds—animals that walk on all fours (although, technically, gorillas walk on their feet and knuckles)

savannas—grasslands with scattered trees and drought-resistant undergrowth such as shrubs

semitropical—referring a climate with hot, humid summers and cooler (but not too cold) winters

species—a group of closely related animals that share a high number of similar genetic traits and are capable of reproducing

stamen—a long, slender tube, usually found in the middle of a flower. The stamen produces and distributes pollen.

subspecies—smaller groups of animals or plants within the larger group known as a species. While subspecies are similar to one another, they have unique qualities that set them apart.

sustainable—grown and/or made in a way that doesn't harm the environment

taboos—things that traditionally are not allowed in a specific community

territorial—protective and defensive of a specific area (territory)

United Nations—a global organization with representation from governments around the world that aims to get countries to work together to solves shared problems

Resources

PRINT

Meltzer, Brad. *I Am Jane Goodall.* Rocky Pond Books, 2016.

Patterson, Francine. *Koko's Kitten*. Scholastic, 1985.

Schrefer, Eliot. *Endangered*. Scholastic, 2012.

Simon, Seymour. *Gorillas*. HarperCollins, 2000.

ONLINE

Ape Initiative: apeinitiative.org/kanzi

Bonobo Conservation Initiative: bonobo.org

Dian Fossey Gorilla Fund: gorillafund.org

Earth Rangers: earthrangers.com

Jane Goodall Institute: janegoodall.org

Orangutan Foundation International: orangutan.org

Ranger Rick: rangerrick.org

Roots & Shoots: rootsandshoots.org

Save the Chimps: savethechimps.org

The Gorilla Foundation: koko.org

UNESCO: Protecting Great Apes and Their Habitats: en.unesco.org/themes/biodiversity/great-apes

Wild for Life: torontozoo.com/tz/podcasts

Wild Kratts: pbskids.org/wildkratts

World Wildlife Fund: worldwildlife.org

Links to external resources are for personal and/or educational use only and are provided in good faith without any express or implied warranty. There is no guarantee given as to the accuracy or currency of any individual item. The author and publisher provide links as a service to readers. This does not imply any endorsement by the author or publisher of any of the content accessed through these links.

FUSE/GETTY IMAGES

Acknowledgments

Thanks to all the great apes at Orca Books, including editor Kirstie Hudson, designer Troy Cunningham and editorial assistant Georgia Bradburne.

Index

Page numbers in **bold** *indicate an image caption.*

CHRISTOPHER GUDGEON writes books, TV shows and movies. He loves animals, nature, ghost stories and magic. His previous books include *Ghost Trackers: The Unreal World of Ghosts, Ghost-Hunting, and the Paranormal* and *Luck of the Draw: True-Life Tales of Lottery Winners and Losers.* He lives in Toronto.